The Absolute Guide to DMT Extraction

Sources, Methods of Extraction, and Potential Therapeutic Benefits

Simon Ashford

Chapter One

Introduction on how to extract dmt

DMT (dimethyltryptamine)
Dimethyltryptamine, or DMT, is a potent hallucinogenic substance that occurs naturally in a wide range of plants, animals, and even the human brain. It is a chemical member of the tryptamine family and shares a structure with serotonin, a neurotransmitter that controls mood, appetite, and sleep. DMT is well known for its strong psychedelic effects, which

frequently result in significant changes to perception, consciousness, and cognition. When consumed, usually by smoking, vaping, or consuming it in drinks such as Ayahuasca, DMT can cause strong aural and visual hallucinations in addition to deep spiritual or mystical experiences. Many people characterize these experiences as transcendent, life-changing, and indescribable.

DMT has been utilized for generations in shamanic and cultural rites because to its potent psychedelic qualities, especially in indigenous populations in South America. DMT has garnered increasing attention in recent years due to its possible therapeutic uses, especially in the treatment of mental health issues and in the study of the nature of consciousness.

But in many nations, including the US, DMT is categorized as a Schedule I controlled substance, which means that without the appropriate permission for research or medical usage, it is unlawful to produce, distribute, or possess the drug. DMT is still being studied despite its illegal status because scientists are interested in how it affects consciousness, the brain, and possible medicinal uses.

The importance of comprehending DMT sources

it's important to comprehend the sources of DMT for a number of reasons:

1. Safety and Harm Reduction: By being aware of the origins of DMT, people are better equipped to choose the substances they take. Users can lower the risk of contamination or adulteration with potentially toxic drugs by being aware of the source of their DMT, ensuring that they are getting it from reliable and safe sources.

2. Cultural and Spiritual Context: DMT has been a component of many cultures' spiritual and cultural traditions for ages, especially those of indigenous people. Gaining knowledge about the origins of DMT helps one to better appreciate and respect these customs by offering insight into the historical and cultural settings in which it has been used.

3. Legal and Ethical Considerations: DMT is categorized as a restricted substance in many nations, and usage, distribution, and manufacture are subject to legal restrictions. Individuals can reduce their risk of legal repercussions by adhering to pertinent legislation and navigating legal frameworks with the assistance of an understanding of the legal status of DMT and its suppliers.

4. Therapeutic Potential: DMT exhibits potential for treating a range of mental health conditions, including addiction, anxiety, and depression. Researchers and medical experts who wish to investigate the therapeutic potential of DMT, including the creation of safe and efficient treatment regimens, must have a thorough understanding of its sources.

5. Neuroscientific Research: Researchers looking into consciousness and the brain processes behind psychedelic experiences are likewise interested in DMT. By conducting trials and investigations to clarify DMT's effects on brain function and subjective experience, researchers can better understand the sources of the drug and further our knowledge of consciousness and altered states of perception.

Chapter Two

DMT's Natural Sources

DMT can be found in nature in a variety of plants, animals, and fungi.

1.Plants:

• **Psychotria Viridis (Chacruna):** One of the main components of Ayahuasca, a hallucinogenic concoction used in customary shamanic ceremonies, is Psychotria viridis, a flowering plant native to the Amazon jungle. There is a lot of DMT in the   leaves.

• **Mimosa Hostilis (Jurema):** Originally from Central and South

America, this tree is also known as Mimosa tenuiflora. Its root bark, which is frequently used to prepare ayahuasca, has a substantial DMT content.

- **Acacia Confusa:** Originally from Southeast Asia, this tree is utilized traditionally in Taiwan for a variety of uses, including the creation of psychoactive concoctions. It contains DMT in its root bark.

2.Animals:

- **Bufo Alvarius, or Colorado River toad:** This species of toad is located in northern Mexico and the southwestern United States. It secretes a compound known as bufotenin, which shares chemical similarities with DMT and when consumed, causes hallucinogenic effects.

• **Different Mammals:** Although its physiological function in these organisms is not entirely understood, DMT has been detected in trace amounts in the brains of humans and other mammals.

3.fungi

• **Psilocybin Mushrooms:** Although the main hallucinogenic ingredient in psilocybin mushrooms is psilocybin, several species also have trace levels of DMT. Because of their hallucinogenic properties, psilocybin mushrooms have been employed in spiritual and shamanic ceremonies for ages. For ages, indigenous tribes have used these natural sources of DMT for shamanic, medical, and spiritual purposes. They are still being researched today because of their possible therapeutic uses as well as their significance in the study of consciousness and altered states of perception. It's crucial to keep in mind, though, that these drugs may be controlled or outlawed in many places, so care should be taken to guarantee responsible and safe use.

Psychotria Viridis (Chacruna)

Psychotria viridis, a perennial shrub usually referred to as Chacruna, is indigenous to the Amazon rainforest, specifically found in Brazil, Peru, and Colombia. It is a member of the Rubiaceae family and is known for having psychedelic qualities because its leaves contain DMT, or dimethyltryptamine.

This is a more thorough synopsis of Psychotria viridis:

1.Synopsis:

• Chacruna usually grows to a height of 3 meters, or around 10 feet, as a shrub. It has elliptical or ovate, glossy green leaves with noticeable veins.

- Clustered at the tips of the stems, Psychotria viridis blooms are small, white, and barely noticeable.

2. Traditional Use:

- Indigenous peoples of the Amazon basin have long used chacruna traditionally, especially in connection with Ayahuasca rites. In these rituals, ayahuasca, a hallucinogenic beverage, is made by combining the leaves of Psychotria viridis with other plants that contain MAO inhibitors.
- An essential ingredient in the Ayahuasca brew, the DMT-containing leaves of Chacruna provide the hallucinogenic effects

that cause visions during ceremonies and rituals.

3. **Chemical Composition**

• Dimethyltryptamine, or DMT, is the main hallucinogenic ingredient of Psychotria viridis. Strong psychedelic drug DMT is well-known for producing hallucinations that frequently result in vivid visual and audio experiences.

• Chacruna leaves also include additional alkaloids, flavonoids, and substances that may be involved in the overall pharmacological effects of DMT.

4. **Cultural Importance:**

• Among the indigenous populations living in the Amazon

basin, chacruna has great cultural and spiritual significance. It is regarded as a sacred plant and plays a major role in many healing rituals and shamanic practices.

● It is said that use chacruna in Ayahuasca ceremonies promotes healing, spiritual discovery, and self-awareness. It's often considered a technique for establishing spiritual connections with nature, the supernatural, and the afterlife.

5. Current Studies and Their Uses:

● Interest in the medicinal properties of Psychotria viridis and its components has grown recently, especially for the treatment of mental health conditions like addiction, depression, and anxiety.

• The pharmacological characteristics, safety profile, and possible therapeutic applications of Chacruna are still being investigated in scientific research. Research endeavors to clarify its impact on the brain, awareness, and mental health.

Hostilis Mimosa (Jurema)

Known by most as Jurema, Mimosa hostilis is a perennial tree or shrub that is indigenous to parts of Central and South America, especially Brazil and Mexico. It is a member of the

Fabaceae family and well-known for both traditional and modern use, such as being a source of the powerful hallucinogenic chemical DMT (Dimethyltryptamine). This is a thorough synopsis of Mimosa hostilis:

1.Description

• Mimosa hostilis usually grows to a height of up to 8 meters (about 26 feet), as a small tree or large shrub. Its leaves are bipinnate, with many tiny leaflets grouped around a central stalk.

• The tree has fragrant, clusters of white to pinkish blooms that open to reveal seed pods filled with seeds.

2. Traditional Use:

• Indigenous peoples of Central and South America, especially Brazil, have long used mimosa hostilis traditionally. It is frequently utilized in many shamanic and spiritual traditions and is highly respected for its hallucinogenic qualities.

The root bark of Mimosa hostilis contains a significant amount of DMT, a powerful psychedelic. Native American tribes have long employed the root bark in hallucinogenic concoctions, frequently mixed with other plants, to promote spiritual insights and hallucinations.

3. Psychoactive Substances:

• Dimethyltryptamine, or DMT, is

the main psychoactive ingredient of Mimosa hostilis. It is a potent psychedelic drug with significant hallucinatory effects on consciousness.

• The root bark of Mimosa hostilis is an excellent source for the extraction of DMT, as it contains the majority of this hallucinogenic chemical.

4. Cultural Importance:

Mimosa hostilis is endemic to Brazil and other places, and among indigenous populations there, it has great cultural and spiritual significance. It is utilized in healing rituals and ceremonial ceremonies and is frequently considered a sacred plant.

• It is said that consuming preparations made from Mimosa hostilis promotes emotional healing, personal development, and spiritual discovery. It is seen as a means of establishing connections with the natural world, the divine, and the collective unconscious.

5. Modern Uses:

• Apart from its customary application, Mimosa hostilis has garnered interest in modern settings, namely in the realms of alternative medicine and psychedelics.

• DMT is frequently extracted from the root bark of Mimosa hostilis, where it is either consumed or subjected to additional processing for a range of uses, including as scientific research, psychotherapy, and spiritual discovery.

Mimosa hostilis, also known as Jurema, is a representation of spirituality, cultural legacy, and the nexus between conventional knowledge and contemporary science. Researchers studying altered states of consciousness and the secrets of the human mind are still captivated by it because of its psychotropic qualities and rich cultural heritage.

Confusa Acacia

Native to Southeast Asia, specifically Taiwan, Acacia confusa is a perennial tree sometimes referred to as Rainbow Tree or Formosa Koa. It is well-known for its many applications and for being a source of the powerful hallucinogenic chemical DMT (Dimethyltryptamine), which it shares with other members of the Fabaceae family. This is a thorough synopsis of Acacia confusa:

1. Description

• Acacia confusa usually grows to be a small to medium-sized tree, with a maximum height of 15 meters, or roughly fifty feet. It features unique bipinnate leaves that resemble ferns and are made up of many tiny leaflets grouped around a central stalk.

• The tree bears seed pods that hold seeds after it produces tiny, yellow blooms that bloom in groups.

2. Traditional and Modern Use:

●The traditional use of Acacia confusa dates back many years in Taiwan, where it was employed for tannin manufacturing, woodworking, and traditional medicinal purposes.

• Acacia confusa has drawn notice lately due to its psychedelic qualities, especially its high DMT concentration. DMT may be extracted from the root bark of Acacia confusa, which is a valuable source of this hallucinogenic substance.

3. Psychoactive Substances:

• Dimethyltryptamine, or DMT, is the main psychoactive ingredient of Acacia confusa. It is a potent psychedelic drug with profoundly hallucinatory effects on consciousness.

• Although minor levels of DMT can also be found in other sections of the Acacia confusa tree, like the leaves and stems, the substance is primarily concentrated in the root bark.

4. Extraction and Ingestion:

• DMT can be extracted from the root bark of Acacia confusa and consumed or processed further for a variety of uses, such as scientific study, psychotherapy, and spiritual discovery.

• To achieve a strong extract, extraction procedures usually entail isolating DMT from the root bark using solvents, then purifying and concentrating the material.

5. Cultural Importance:

• Although Acacia confusa is not as culturally significant as other plants that contain DMT, such as Psychotria viridis or Mimosa hostilis, it has still gained prominence in the psychedelic community's toolkit.

• Similar to other DMT-containing plants used in traditional shamanic rituals, Acacia confusa preparations are said to cause visionary experiences, spiritual insights, and altered states of consciousness upon consumption.

Chapter Three

Animals that have DMT in them

Although DMT (dimethyltryptamine) is well recognized for being present in a variety of plant species, it can also be found, albeit in smaller amounts, in some mammals.

1. The Colorado River toad, Bufo Alvarius:

● The Colorado River toad, sometimes referred to as the Sonoran Desert toad or Bufo alvarius, secretes a chemically comparable compound called bufotenin. When consumed, the psychoactive substance bufotenin causes hallucinations.

• Although bufotenin and DMT are not exactly the same, they are related in terms of their chemical makeup and psychoactive characteristics. The toad's skin can produce strong psychedelic effects when it is dried and smoked.

2. Different Mammals

• Traces of DMT have been discovered in the brains of humans and other mammals. Although DMT's exact function in these creatures is unknown, its presence points to the possibility that it serves some physiological purpose.

• Scientists are currently

investigating whether endogenous DMT production occurs in animals, and they have hypothesized that DMT may have a role in consciousness, dreaming, and near-death experiences.

3. Additional Invertebrates and Amphibians:

• Although the most well-known animal known to possess bufotenin is the Bufo alvarius toad, other toad and amphibian species may also manufacture this substance to variable degrees.

• Moreover, there is evidence that suggests several invertebrate species, including some marine sponge and octopus species, may contain DMT or similar substances.

It's crucial to remember that even while these creatures might contain psychoactive substances like DMT or bufotenin, using them recreationally or for spiritual reasons may present ethical and legal challenges. Animals may suffer injury or decline in natural populations if their psychotropic secretions are harvested or used for commercial purposes. Therefore, while considering the use of animals for psychedelic experiences, ethical and acceptable practices should be followed.

Colorado River toad, Bufo Alvarius, contains bufotenin.

Bufotenin, sometimes referred to as 5-HO-DMT (5-hydroxy-N,N-dimethyltryptamine), is a hallucinogenic substance that is present in the venom of the Bufo alvarius toad, which is also referred to as the Sonoran Desert or Colorado River toad. This is a thorough summary of bufotenin in Bufo alvarius:

1. Natural Environment:

• The Bufo alvarius toad is indigenous to the Sonoran Desert and other desert regions in northern Mexico and the southwestern United States. Its squat body, warty skin, and enormous venom-producing glands behind its eyes give it a unique appearance.

2. The composition of venom:

• Among the several bioactive substances found in the venom of the Bufo alvarius toad is bufotenin. Bufotenin is an alkaloid of tryptamine that shares chemical similarities with DMT (dimethyltryptamine). Being a serotonin analog, it is structurally identical to serotonin, a neurotransmitter that regulates mood.

3. Effects on the mind:

• A strong psychoactive substance called butogenenin is well-known for producing hallucinations when consumed or smoked. By attaching to serotonin receptors in the brain and inducing altered states of consciousness, it primarily functions as a serotonin receptor agonist.

• The dried and smoked venom of Bufo alvarius can cause strong psychedelic experiences that include visual hallucinations, changed perceptions of time and space, and significant mood and cognitive shifts.

4. Spiritual and Cultural Use:

• For ceremonial and spiritual purposes, indigenous peoples of the southwestern United States and northern Mexico have traditionally employed the venom of the Bufo alvarius toad. It is thought that bufotenin's psychotropic effects promote spiritual inquiry, healing, and enlightenment.

• The usage of Bufo alvarius toad venom has become more common in modern times among people who are interested in spiritual development and psychedelic experiences. In therapeutic or guided ceremonies, some practitioners employ the venom.

5. A Legal and Ethical Perspective:

Depending on the jurisdiction, Bufo alvarius toad venom has a different legal status. Legal limitations may apply in some places to the extraction and use of toad venom for spiritual or recreational purposes.

• There are also moral issues with the collection and application of toad venom. To protect wild populations as little as possible and to guarantee the well-being of the toads, sustainable and compassionate methods should be used.

DMT in human beings and other mammals

Dimethyltryptamine, or DMT, has been detected at trace levels in a number of species, including humans. An outline of DMT's occurrence in mammals is as follows:

1. Internalized DMT:

• DMT that is created by the body spontaneously is referred to as endogenous DMT. Although the precise biological mechanisms and roles of endogenous do not completely understand, research indicates that certain tissues and organs, including the brain, generate modest amounts of it.

• Research on people has found that biological fluids such blood, urine, and cerebrospinal fluid contain DMT and its metabolites. On the other hand, endogenous DMT concentrations are usually far lower than those in plants or artificial preparations.

2. Biological Purposes:

• Research and discussion on the potential biological roles of endogenous DMT in mammals, including humans, are still underway. According to some research, DMT may be involved in immunological response, neurotransmission, and neuroprotection, among other physiological functions.

- It has also been suggested that DMT plays a role in the control of mood, consciousness, and cognition. Endogenous DMT has been proposed by some scientists as a potential factor in mystical or spiritual experiences and spontaneous altered states of consciousness.

3. The metabolic pathways

- It is thought that a sequence of biochemical processes including enzymes like indolethylamine-Nmethyltransferase (INMT) manufacture DMT from the amino acid tryptophan. Given that INMT is present in the brain, lungs, and liver, among other tissues, it is possible that numerous organs manufacture DMT.

- After production, enzymes including aldehyde dehydrogenase (ALDH) and monoamine oxidase (MAO) quickly break down DMT, forming inactive metabolites such 5-hydroxyindole-3-acetic acid (5-HIAA) and indole-3-acetic acid (IAA).

4. Distribution Biologically:

● Although DMT and its metabolites have been found in a variety of physiological fluids and tissues, it is still unclear exactly where endogenous DMT is found in the body and how it is distributed. DMT has been linked to melatonin synthesis and circadian rhythms, and research indicates that it may be found in particular brain areas, including the pineal gland.

Chapter Four

Fungi that have DMT in them

Compared to plants and animals, fungi are less frequently mentioned when it comes to DMT, however several species have been reported to possess this psychedelic substance. This is a synopsis:

1. Psychedelic Mushrooms:

• Although the main hallucinogenic ingredients of psilocybin mushrooms are psilocybin and psilocin, several species of the Psilocybe genus have also been shown to contain trace levels of DMT.

- Along with psilocybin and psilocin, psilocybe cubensis, one of the most popularly grown and consumed species of hallucinogenic mushrooms, has also been confirmed to contain trace amounts of DMT.

2. Other Types of Mushrooms:

- Although little research has been done, anecdotal evidence and scant scientific data point to the possibility that other mushroom species contain DMT or similar substances.

- For instance, it has been proposed that certain Gymnopilus species contain DMT, although more investigation is required to support these claims.

3. Mycoflora and Knowledge of Ethnobotany:

• Fungi have long been used in spiritual, therapeutic, and ceremonial contexts by traditional and indigenous civilizations. Although there may not be much specific knowledge about fungi that contain DMT, fungi in general, including mushrooms, have been significant components of cultural rituals and ethnomycological traditions.

• Although there may be little evidence of these traditions, some societies may have used mushrooms containing DMT or other psychoactive substances for shamanic rituals, healing ceremonies, and spiritual discovery.

4. Research and Identification:

Comparing the study of DMT-containing fungi to other sources, such as plants and animals, is still in its infancy. With advancements in taxonomic categorization and analytical methods, research endeavors to identify and define psychotropic chemicals in fungi are ongoing.

• With millions of species thought to exist, fungi offer enormous promise for the discovery of novel chemicals and the comprehension of their medicinal and ecological implications.

Psilocybin mushrooms and DMT: a connection

Although both psilocybin mushrooms and DMT (dimethyltryptamine) are psychedelics, their biological origins, effects, and chemical makeup are different. This is a summary of their relationship:

1. Chemical composition:

• The principal hallucinogenic ingredients of psilocybin mushrooms are psilocybin and psilocin. As a pro-drug of psilocin, psilocybin is transformed by the body into psilocin, which then binds to serotonin receptors in the brain to generate psychedelic effects.

• DMT is a distinct molecule that is a part of the tryptamine family. While its effects as a serotonin receptor agonist are somewhat comparable to those of psilocybin and psilocin, they are chemically different from each other.

2. Natural Sources

• Psilocybin mushrooms, commonly referred to as "**magic mushrooms,**" are a type of fungus that fall under, several genera, such as Panaeolus and Psilocybe. Psilocybin and psilocin are naturally produced by these mushrooms through metabolic mechanisms.

- DMT is present in a wide range of flora, fauna, and fungi. Although the DMT content of psilocybin mushrooms is negligible, several species of Psilocybe have been shown to have trace levels of DMT in addition to psilocybin and psilocin.

3. Impacts:

- Changes in perception, mood, and cognition are hallmarks of the psychedelic effects produced by psilocybin mushrooms and DMT. However, other factors including dosage, individual characteristics, and environmental setting can affect the exact type and duration of these effects.

• When smoked or vaporized, psilocybin mushrooms usually result in a longer-lasting and more contemplative experience than DMT, which is recognized for its potent and comparatively fleeting effects, sometimes lasting 15 to 30 minutes.

4. Potential for Therapeutic Effects:

• Research has demonstrated the potential of psilocybin mushrooms and DMT as therapeutic agents for the treatment of a range of mental health conditions, such as anxiety, depression, and post-traumatic stress disorder (PTSD).

- Studies on the therapeutic potential of psilocybin-assisted treatment have shown promise in symptom relief and enhancement of psychological health. Analogously, early research on DMT's therapeutic application points to the drug's possible effectiveness in treating diseases including addiction and depression.

5. Legal Status:

• The euphoric properties and potential for misuse of DMT and psilocybin mushrooms have led to their classification as prohibited substances in numerous nations. Nevertheless, each jurisdiction has a different legal position regarding these narcotics; some decriminalize or allow their use under specific conditions.

• Although DMT and psilocybin mushrooms are both hallucinogenic compounds with potential for therapeutic use and overlapping effects, their chemical makeup, suppliers, and legal status are different. To completely comprehend their modes of action, therapeutic uses, and possible hazards, more investigation is required.

Chapter Five

DMT production in a lab

Dimethyltryptamine, or DMT, can be produced in a lab setting by following a few rather straightforward chemical steps. This is a summary of how DMT is synthesized in a lab:

1. first Resources:

• Tryptamine is a naturally occurring substance that is present in a wide variety of creatures, including plants, animals, and fungi. It serves as the main precursor for the creation of DMT. The primary building block from which DMT is generated is tryptamine.

2. Chemical Processes:

• Tryptamine reacts with an alkyl halide, like methyl iodide or ethyl bromide, in the presence of a base, like potassium carbonate (K_2CO_3) or sodium hydroxide (NaOH), to produce DMT. This reaction is a typical way to synthesize DMT. Tryptamine's amino group is alkylated as a result of this process, resulting in the formation of N,N-dialkyltryptamine derivatives, which include DMT.

• Tryptamine can also be synthesized by condensation with

formamide or formic acid in an acidic environment. N,N-dialkyltryptamine derivatives are produced by this reaction, and these can be processed further to produce DMT.

3. The purification process:

• To isolate DMT in its pure form, the crude product from the synthesis step is usually purified using methods like chromatography or re-crystallization. Purification ensures the quality and purity of the finished product by assisting in the removal of impurities and by-products produced during the synthesis process.

4. Characterization

• After being refined, the synthesized DMT is examined utilizing analytical methods to verify its identification and structural integrity, including mass spectrometry (MS), infrared (IR) spectroscopy, and nuclear magnetic resonance (NMR) spectroscopy.

5. Handling and Storing:

• To avoid deterioration, synthesized DMT is normally kept in sealed containers away from light and moisture. To protect people and reduce exposure to the material, proper handling techniques must be followed.

The chemical method used to create DMT

Dimethyltryptamine (DMT) is made chemically by a number of stages that begin with basic raw materials. Alkylating tryptamine with an alkyl halide is one popular technique. This is a condensed synopsis of the chemical process:

1. Starting Resources:

• Tryptamine: Tryptamine is a naturally occurring substance that is essential to the synthesis of DMT.

• Alkyl Halide: To add alkyl groups to the nitrogen atom of tryptamine, an alkyl halide, such as methyl iodide (CH3I) or ethyl bromide (C2H5Br), is utilized as the alkylating agent.

2. Alkylation Process:

• Tryptamine is dissolved in a suitable solvent in a reaction vessel, such as dimethylformamide (DMF) or anhydrous ethanol.

• To stop oxidation, the alkyl halide typically methyl iodide is added drop wise to the tryptamine solution while it is enclosed in an inert environment, such as nitrogen or argon.

• To facilitate the alkylation reaction, the reaction mixture is heated to reflux temperature, or between 80 and 100°C, and agitated for a number of hours. Tryptamine's amino group interacts with the alkyl halide to generate N,N-dialkyltryptamine derivatives, which include DMT.

3. Work-Up and Cleaning:

• Following the completion of the reaction, the reaction mixture is cooled, and the crude product is extracted or evaporated using a solvent.

• After that, contaminants are eliminated from the crude product and pure DMT is isolated using methods including re-crystallization and column chromatography.

4. Characterization

• To verify the identity and structural integrity of the purified DMT product, analytical methods including mass spectrometry (MS), infrared (IR) spectroscopy, and nuclear magnetic resonance (NMR) spectroscopy are employed.

5. Handling and storing process:

• To avoid deterioration, the synthesized DMT is kept dry and sealed in containers away from light and moisture. To protect people and reduce exposure to the material, proper handling techniques must be followed.

Ethical and legal aspects

DMT has many complex legal and ethical implications that need to be properly considered. Below is a summary of some important points:

1. Status Legal:

- DMT is regarded as having a high potential for abuse and no recognized medicinal value, making it a Schedule I controlled substance in many nations, including the US. DMT distribution, use, and possession outside of approved medical or research settings are prohibited and may result in harsh legal repercussions, such as fines and jail time.

• To prevent legal ramifications, it is imperative that you are informed of the particular laws and rules that regulate DMT in your area and that you abide by them.

2. Safety and Mitigation of Injury:

• There are inherent hazards associated with the synthesis, possession, and use of DMT, including the possibility of psychological and physical injury.

Working with dangerous chemicals that provide risks to health and safety if improperly handled is a necessary part of the field of synthetic chemistry. Furthermore, the strong psychoactive effects of DMT may result in uncomfortable sensations as well as altered states of consciousness.

• Safety and harm reduction measures, like appropriate laboratory techniques, personal protective equipment, and informed consent, should be given top priority by practitioners. DMT users should also be knowledgeable about the drug's effects, its hazards, and ways to reduce them.

3. Usage in Ethics:

• Respect for indigenous cultures and customs, informed consent, and responsible usage are only a few of the many ethical issues surrounding DMT. Because of their cultural and spiritual linkages to plants that contain DMT, indigenous tribes in areas where these plants are traditionally used should have their knowledge and customs respected and preserved.

• When using DMT for therapeutic purposes or research, practitioners must get participants' informed consent. This entails giving participant's complete information about the study's objectives, any drawbacks and advantages, confidentiality, and their right to discontinue participation at any moment.

• Using DMT responsibly entails consuming it with mindfulness, reverence, and intention. The integration of experiences, mental readiness, and set and setting are crucial components of moral DMT use.

4. Investigation and Therapeutic Applications:

• Investigations into the mechanisms of action and

therapeutic possibilities of DMT are still underway. It is crucial to follow ethical norms and regulations, adhere to strict scientific methods, and ensure participant safety when conducting research.

• Established ethical standards, such as beneficence, non-malfeasance, autonomy, and fairness, should be followed in DMT-assisted therapy and research. Research ought to be carried out with honesty, openness, and regard for human rights and dignity, carefully balancing any potential advantages and hazards.

Chapter Six

Techniques for extracting from natural sources

There are various techniques used to extract DMT from natural sources, such as plants that naturally contain the chemical. Here are a few popular methods of extraction:

1. Acid-Base Extraction:

One of the most popular techniques for removing DMT from plant materials is acid-base extraction. It includes separating DMT from other plant components by utilizing basic and acidic solutions.

- To change the DMT salts into their freebase form, which is more soluble in nonpolar solvents, the plant material, which is usually dried and powdered, is first soaked in an acidic solution, like citric acid or vinegar

- To dissolve the DMT freebase, the plant material is extracted using a nonpolar solvent, such as petroleum ether or naphtha, following acidification. To obtain optimal yield, this extraction procedure is usually performed several times.

- After extracting the nonpolar solvent containing the DMT

freebase from the acidic aqueous layer, it is evaporated to produce crude DMT, which is optionally further refined.

2. Steam Distillation Process:

Another technique for removing DMT from plant materials is steam distillation, which is especially useful when the substance is found in essential oils.

• When steam is passed over plant material, DMT-containing essential oils vaporize. This process is known as mist cleansing. After condensing and gathering the vapor, an essential oil mixture is produced.

• To separate DMT from other components, the essential oil mixture is subsequently put through additional purification processes including solvent extraction or chromatography.

3. Extracting Solvents Process:

• Solvent extraction is the process of removing DMT from plant materials using organic solvents such ethanol, methanol, or acetone.

• In order to aid in the extraction of DMT and other chemicals, the plant material is usually macerated or powdered and then soaked in the solvent.

• Following the soaking process, the plant material is filtered out of the solvent, and the solvent containing the extracted DMT is then evaporated at a lower pressure to produce crude DMT extract.

• Pure DMT can be separated from the crude extract by employing methods like re-crystallization or chromatography.

4. Extracting with Cold Water:

• A straightforward technique for removing DMT from plant sources containing water-soluble DMT salts is cold water extraction.

- To dissolve the DMT salts, the plant material is steeped in cold water for a long time.

- Following soaking, the plant material is removed from the water by filtering it, and the DMT-containing solution is then evaporated under low pressure to produce crude DMT extract.

The crude extract can be refined even more by employing methods like chromatography or re-crystallization.

5. Using Enzymes for Extraction:

- Enzymatic extraction releases DMT and other chemicals by

breaking down plant cell walls with the help of enzymes.

• In order to facilitate extraction, the plant material is usually combined with enzymes, such as cellulase or pectinase, and incubated under carefully regulated conditions.

• The mixture is filtered to eliminate solid residues following the enzymatic treatment, and the DMT-containing solution is then processed one more time to isolate and purify DMT.

Chapter Seven

Methods for removing DMT from Plants

Dimethyltryptamine, or DMT, is usually extracted from plants using a variety of methods to separate the molecule from the plant material. Here are a few popular techniques:

1. Base-Acid Extraction:

The technique of acid-base extraction is commonly employed to extract DMT from plant material. This procedure makes use of DMT's weak base and ability to change into its more soluble freebase form in nonpolar liquids.

- To change DMT salts into the freebase form, the plant material—which is typically dried and powdered must first be soaked in an acidic solution, like vinegar or citric acid.

- To dissolve the DMT freebase, the plant material is extracted using a nonpolar solvent, such as petroleum ether or naphtha, following acidification.

- To obtain crude DMT, which can optionally be further refined, the nonpolar solvent layer containing the DMT is separated from the acidic aqueous layer and then evaporated.

2. Steam Distillation Process:

- One popular technique for extracting essential oils from plant material, especially DMT-containing plants, is steam distillation.

- When steam is passed over plant material, DMT-containing essential oils vaporize. This procedure is called steam distillation.

- After condensing and gathering the vapor, an essential oil blend containing DMT is produced.

- To separate DMT from other components, the essential oil mixture might be subjected to additional processing methods such solvent extraction or chromatography.

3. Extracting Solvents:

• Solvent extraction is the process of removing DMT from plant material using organic solvents such ethanol, methanol, or acetone.

• In order to aid in the extraction of DMT and other chemicals, the plant material is usually macerated or powdered and then soaked in the solvent.

• Following soaking, the plant material is removed from the solvent by filtering it, and the solvent that contained the extracted DMT is then evaporated to produce crude DMT extract.

4. Extracting with Cold Water:

• Water-soluble DMT salts can be extracted from plant material using a straightforward technique called cold water extraction.

• To dissolve the DMT salts, the plant material is steeped in cold water for a long time.

• Following the soaking period, the plant material-containing water is filtered out, and the DMT-containing solution is evaporated to produce crude DMT extract.

These are only a handful of the methods utilized to extract DMT from plants. The concentration of DMT, the desired level of purity for the finished product, and the chemical characteristics of the plant material all influence the extraction technique selection. It's crucial to remember that DMT extraction from plants needs to be done carefully and in accordance with safety and legal regulations.

DMT's potential for therapy

When compared to other psychedelics like psilocybin and LSD, research on the therapeutic potential of DMT (dimethyltryptamine) has been very limited. However, interest in this field has grown in recent years. Despite this, research on the therapeutic effects of DMT is still ongoing and is gaining popularity. The following are a few possible medicinal uses for DMT:

1. Impact of Psychotherapy:

• DMT has been shown to cause significant changes in consciousness, such as ego dissolution, mystical-type experiences, and a sense of oneness with the cosmos. In psychotherapy, these experiences might be therapeutically valuable, especially for those who are dealing with addiction, despair, anxiety, and existential anguish.

• According to some researchers, DMT-assisted therapy may improve insight, ease the processing of emotions, foster psychological development, and improve overall wellbeing. When DMT experiences are included into a therapy setting, people may be able to see their lives, relationships, and existential issues from fresh angles.

2. Management of Anxiety and Depression:

• Anecdotal accounts and preliminary research point to

possible anxiolytic and depressive effects of DMT. There is evidence linking DMT-induced mystical experiences to decreased anxiety, existential discomfort, and depressive symptoms.

- Studies into DMT's possible mechanisms of action and therapeutic efficacy in clinical populations are being conducted in an effort to better understand the drug's antidepressant effects.

3. Treatment for Addiction:

- DMT-assisted therapy has demonstrated promise as a possible treatment for addictions related to substance use, such as alcohol and tobacco addiction.

Mystical experiences brought on by DMT may help people overcome bad behavioral patterns, develop a feeling of purpose and connection, and obtain insight into the fundamental causes of their addiction.

• Research has indicated that DMT may promote neuroplasticity and interfere with dysfunctional brain circuits linked to addiction, resulting in long-term alterations in behavior and thought processes.

4. Investigating the Spiritual and Existential:

- Indigenous and shamanic traditions have long used DMT for spiritual and therapeutic purposes. People may gain understanding of the nature of reality, awareness, and the self through the profoundly altered states of consciousness brought on by DMT.

- Visionary visuals, mystical creatures, and transcendental states of consciousness are frequently encountered during DMT experiences. One's worldview and sense of meaning and purpose may be profoundly altered by these encounters, which can evoke sentiments of awe, veneration, and oneness.

5.   Effects   on   the Neurobiology:

● Although studies on DMT's neurobiological effects are still in their early phases, it has been proposed that the drug may alter brain serotonin receptors, namely the 5-HT2A receptor subtype. Since serotonin signaling is essential for mood control, thought processes, and processing emotions, changes in serotonin transmission brought on by DMT may be part of the drug's therapeutic benefits.

Chapter Eight

Assisted psychedelic treatment

A therapeutic strategy known as "psychedelic-assisted therapy" includes the use of psychedelic drugs as supplements to psychotherapy, such as psilocybin, MDMA, or DMT.

1. Fundamentals:

● A systematic treatment approach that combines the use of psychedelic drugs with encouraging psychotherapy techniques is usually involved in psychedelic-assisted therapy.

• Trained therapists or facilitators lead the therapy and offer a secure and encouraging environment for the person going through the process.

• Preparation, the psychedelic drug's administration, the peak psychedelic experience, and integration thereafter are frequently included in sessions.

2. Psychoactive Drugs:

• Clinical studies have demonstrated the therapeutic potential of a number of psychedelic drugs. These include ketamine, LSD (lysergic acid diethylamide), MDMA (3,4-methylenedioxymethamphetamin e), psilocybin (found in some species of mushrooms), and ayahuasca (which contains DMT).

• Although every substance is different in its effects and modes of action, they all have some things in common, like encouraging emotional transparency, strengthening introspection, and enabling significant alterations in awareness.

3. Uses:

• A number of mental health issues, such as depression, anxiety, PTSD, addiction, and end-of-life misery, have been studied in relation to psychedelic-assisted therapy.

• According to research, psychedelics may facilitate the processing of challenging emotions, provide insight into underlying psychological problems, and enable people to break free from constrictive thought and behavior patterns.

4. Mechanisms of Action:

• It is believed that psychedelics' therapeutic benefits are mediated by their interactions with the brain's serotonin receptors, specifically the 5-HT2A receptor subtype. Psychedelics have the potential to alter brain circuitry related to emotion processing, thought, and mood control.

● Moreover, psychedelics may enhance neuroplasticity, or the brain's capacity to rearrange and create new connections, which may be the basis for their therapeutic benefits over an extended period of time.

5. Risks and Safety:

● To reduce hazards and optimize therapeutic benefits, psychedelic-assisted treatment is administered in a highly regulated setting.

● Although psychedelics are usually regarded as safe when

used in therapeutic settings, some people may encounter powerful psychological experiences or have difficult or upsetting reactions as a result of using them.

Psychedelic side effects might include temporary elevations in heart rate and blood pressure as well as anxiety, paranoia, and bewilderment. On the other hand, when therapy is administered under the proper supervision, major side effects are uncommon.

6. Legal and Regulatory Factors to Consider:

• Psychedelics' legal standing differs depending on the

jurisdiction; some are considered controlled substances, while others are being studied for possible medical applications.

In certain areas, there is a growing movement to decriminalize or legalize psychedelics for therapeutic purposes due to a change in public opinion and growing awareness of their potential advantages.

Chapter Nine

DMT's effects on the brain

Dimethyltryptamine, or DMT, has a wide range of complicated effects on the brain that can affect cognition, emotion, perception, awareness, and behavior. Research points to a number of possible ways that DMT interacts with the brain, even if our knowledge of the precise mechanisms underpinning its effects is still developing.

1. Agonism of the Serotonin Receptor:

● DMT mainly functions as an agonist of serotonin receptors, specifically those of the 5-HT2A subtype. The brain is filled with serotonin receptors, which play a major role in controlling mood, perception, thought processes, and several physiological functions.

- It is thought that DMT's activation of 5-HT2A receptors modifies brain activity in cortical and subcortical regions, resulting in changes to consciousness and perception typical of the DMT experience.

2. Enhanced Connectivity in the Brain:

- DMT administration has been linked to increased functional connectivity between brain regions, particularly in areas related to higher-order cognition, sensory processing, and self-awareness, according to neuroimaging studies using methods like functional magnetic resonance imaging (fMRI).

- Many users have reported subjective experiences of ego dissolution, unity consciousness, and a sense of interconnectedness with the universe, which may be related to DMT-induced changes in brain connectivity.

3. Default Mode Network (DMN) Disruption:

• During rest and self-referential thought, the default mode network (DMN) is a network of brain regions that are active. Alterations in consciousness and a number of psychiatric disorders have been linked to disruptions in the dorsal mediaeval nervous system.

• It has been demonstrated that administering DMT temporarily disrupts the DMN's activity and connectivity, which alters self-referential processing, introspection, and ego boundaries.

4. Neuroplasticity and Remodeling of Synapses:

• By modifying synaptic signaling and gene expression, psychedelics like DMT may enhance neuroplasticity the brain's capacity to rearrange and form new connections—according to some research. This could be a factor in the psychedelics' long-lasting therapeutic benefits that have been shown in clinical research.

• Research on animals has demonstrated that the administration of DMT can cause structural alterations in neurons, such as dendritic growth, spine formation, and synaptogenesis, especially in areas of the brain linked to memory and learning.

5. Subcortical Structure Activation:

• It has been demonstrated that the administration of DMT activates subcortical structures related to sensory processing, emotion regulation, and memory formation, including the thalamus, amygdala, and hippocampus.

These subcortical structures may be activated during DMT trips, which could explain the strong emotions, vivid imagery, and altered perception of time and space.

Therapy for mental health issues

1. Psychoanalysis:

Psychotherapy, commonly referred to as talk therapy or counseling, is an essential part of mental health care. It entails having sessions with a licensed therapist to discuss ideas, emotions, and actions as well as to create coping mechanisms and problem-solving techniques.

• Depending on the diagnosis and preferences of the patient, several forms of psychotherapy may be employed, such as dialectical behavior therapy (DBT), cognitive-behavioral therapy (CBT), interpersonal therapy (IPT), and psychodynamic therapy.

2. Drugs:

• Drugs are frequently prescribed to treat mental health conditions, especially attention-deficit/hyperactivity disorder (ADHD), bipolar disorder, depression, anxiety, and schizophrenia.

• Antidepressants, anxiety reducers, mood stabilizers, antipsychotics, and stimulants are among the frequently prescribed drugs. Working closely with a healthcare professional is crucial to selecting the appropriate medication and dosage and to keep an eye out for any possible side effects.

3. Changes in Lifestyle:

- Lifestyle modifications can enhance other forms of treatment because they have a substantial impact on mental health. This can entail consistent physical activity, a healthy diet, enough rest, stress-reduction methods (like mindfulness, meditation, and yoga), and abstaining from substances like alcohol and drugs that exacerbate symptoms.

- Taking part in fulfilling hobbies, social interactions, and meaningful activities can all help to support mental health and give one a feeling of fulfillment and purpose.

4. Assistive Services:

• People with mental health disorders can benefit greatly from the support and encouragement that supportive services and community resources can offer. Support groups, peer counseling, case management, housing assistance, vocational rehabilitation, and crisis intervention services are a few examples of this.

• Establishing a solid support system with family, friends, and medical professionals can help during trying times by providing emotional support, useful help, and encouragement.

5. Alternative and Complementary Medicines:

• Complementary and alternative therapies, such as massage therapy, art therapy, music therapy, acupuncture, and animal-assisted therapy, may provide some people with relief from mental health symptoms.

• Although these methods might not be a replacement for research-backed therapies, they can supplement conventional counseling and medicine and offer more opportunities for self-expression, unwinding, and healing.

6. Using Psychedelics in Therapy:

• As previously noted, psychedelic-assisted therapy is a new therapeutic approach that has promise for treating a variety of mental health conditions, such as addiction, PTSD, anxiety, and depression. It entails the therapeutically supervised use of psychedelics to promote profound emotional processing, awakening, and personal development.

Chapter Ten

Safety precautions for using DMT

It is imperative that DMT users follow safety precautions in order to protect their health and well-being when using this potent psychedelic. Although DMT is generally regarded as safe when used sensibly and in the right situations, it can cause strong psychological effects and carry risks, especially when used carelessly or unsupervised.

Configuration and Environment:

• Select a familiar, secure, and comfortable setting for ingesting DMT; ideally, this should be a calm, quiet area free from hazards and distractions.

• Before taking DMT, be mindful of your emotional state and mentality. DMT should not be taken when you are experiencing severe anxiety, stress, or unstable emotions.

• Be in the company of dependable and encouraging people who can offer consolation, direction, and help when required.

2. Getting ready:

• Before taking DMT, familiarize yourself with its effects, risks, and possible advantages. Recognize that DMT can cause strong, sometimes overwhelming, changes in consciousness and perception.

• Take into account your goals and reasons for utilizing DMT. Be open-minded, curious, and respectful of the strength and potency of the substance as you approach the experience.

• In the days preceding the DMT experience, make sure you eat a healthy diet and stay hydrated. Steer clear of anything that could counteract the effects of DMT, such as caffeine, alcohol, or other substances.

3. Administration and Dosage:

• If you are new to DMT or are not familiar with its effects, start with a low to moderate dosage. As necessary, gradually raise the dosage until the desired degree of intensity is reached.

• Choose a method of administration that suits your preferences and experience level, such as vaporization, smoking, or oral ingestion (e.g., ayahuasca). Use reliable and accurate dosing equipment to measure the amount of DMT consumed.

• Be aware that the onset of DMT effects is rapid, typically occurring within seconds to minutes after administration, and the duration of the experience is relatively short, usually lasting around 15 to 30 minutes.

4. Safety Precautions:

• Prioritize physical safety during the DMT experience. Ensure that the consumption area is free from sharp objects, open flames, or other potential hazards.

• Consider using a sitter or trip sitter—a trusted individual who remains sober and provides support and guidance throughout the DMT experience. The sitter can assist with grounding techniques, reassurance, and ensuring the safety of the individual.

• If using DMT in a group setting, establish clear communication and boundaries with other participants. Respect each other's space and autonomy, and avoid engaging in risky or harmful behaviors.

5. Integration:

• After the DMT experience, take time to reflect on the insights, emotions, and perceptions that arose during the trip. Journaling, meditation, or discussions with a therapist or trusted confidant can aid in processing and integrating the experience.

• Be patient and gentle with yourself as you integrate the lessons and insights gained from the DMT experience into your daily life. Allow time for reflection, growth, and self-discovery.

Conclusion

In conclusion, DMT (Dimethyltryptamine) is a powerful psychedelic compound that induces profound alterations in consciousness and perception. Whether sourced from natural substances like certain plants or synthesized in a laboratory, DMT has garnered significant interest for its potential therapeutic applications, spiritual exploration, and scientific inquiry.

Understanding the sources of DMT, including plants like Psychotria viridis, Mimosa hostilis, and Acacia confusa, as well as animals like the Bufo alvarius toad and various mammals, provides insights into the diversity of natural sources containing this compound.

Extraction methods from these sources involve techniques such as acid-base extraction, steam distillation, solvent extraction, and cold water extraction, each tailored to isolate DMT effectively.

The therapeutic potential of DMT spans a wide range of mental health disorders, including depression, anxiety, addiction, and PTSD. Psychedelic-assisted therapy, which incorporates the use of DMT in therapeutic contexts, offers promising avenues for healing and personal growth, though it requires careful consideration of safety, legality, and ethical guidelines.

When consuming DMT, whether for therapeutic or recreational purposes, it is essential to adhere to safety guidelines, including dosage awareness, setting, and mental preparedness. Additionally, understanding legal and ethical considerations surrounding DMT use is crucial to ensure responsible and informed engagement with this potent psychedelic substance.